INTRODUCING

the QR Code:

The Reality & the Magic

Judith Sansweet

TABLE OF CONTENTS

Introducing the QR Code

In our increasingly mobile world, 90% of the world's population will soon be within the coverage of wireless networks, and there are already an estimated five **billion** cell phones used globally.

Now, this ubiquitous device has sprouted a new dimension; let's call it the QR Revolution.

You've probably seen them and not known what they are. You may even have heard people talking about them – or talked about them yourself. Although they have been in wide use in Japan, other Asian countries, and Europe for the past five years, most people in the Western world: especially the USA, UK, Australia, and New Zealand, are just beginning to wonder, **What are they?**

1

What is a QR Code?

A QR Code (QR=Quick Response) is a two-dimensional matrix code which is essentially a two-dimensional barcode (2D) created by the Japanese firm Denso Wave in 1994. The QR Codes remained largely unused until recently (2006) when advancing technology and the exponential growth of mobile telephony and communications began to bring them into the public marketplace.

QR Codes are matrix-type codes developed to accomplish three primary goals:

- to enable fast, multi-directional scanning
- to provide high capacity storage within a small area
- to put unlimited portable information at your fingertips.

How are these QR Codes used?

QR Codes are primarily meant to be scanned by a mobile device. Anyone with a late model Nokia N series or similar phone is likely to have a barcode reader already installed. Most people with iPhones can easily install the readers, and many of the Japanese brand mobile phones come with inbuilt QR Code readers.

Using them is as simple as 1-2-3:

- Scan the code with the camera of your mobile phone.
- See results.
- Open the connecting link, or save results.

How do they work?

QR codes can be read by the camera in most mobile phones. Once the code is scanned, software in the phone deciphers the encrypted information and initiates a response. This response can involve either directing the phone to a website, providing the user with information stored within the code, or instigating an event. In any case, it eliminates the need to search for content, and provides immediate, direct access to information.

What are they used for?

- Advertising
- Competitions
- Business cards
- Social networking (Facebook, Twitter, etc.)
- Branding
- Ticketing/registration
- Campaign tracking
- File access
- Statistics
- Logistics/parts tracking
- Marketing
- Payment
- Reminders/updates
- the possibilities are endless.

 In fact, Louis Vuitton and Takashi Murakami have created a designer QR code just for the beauty of it!

Where are the QR Codes being used?

They are most actively used in Japan, Hong Kong, Singapore, China, Europe, and Scandinavia — so far.

Are you familiar with your local Entertainment coupon book? In Singapore, they have their own version, full of QR Codes! In these countries, most popular smart phones now come with the QR Code reader already installed. Almost all other smart phones will support the technology required to install a QR Code reader. QR Code is currently being used to promote a British film, *The British Police*.

Who uses QR Codes?

The QR Code is standardized; this shows that QR Code has been accepted internationally, and ensures its widespread global adoption. It can be used domestically as well as internationally.

At the moment, it is mostly used by young audiences; i.e. Gen X and Y, but we can expect this demographic to grow. The international market has really taken hold of these codes. Huge billboard advertising that incorporates QR codes is taking place across Asia and Europe. Usage in Japan is approaching 100%.

In 2008, The *Good Morning America* TV show ran this as an American intro: Check it out on YouTube. http://www.youtube.com/watch?v=5RFQ0cnS Ohg

For a more current picture, see

http://www.qrmonkey.com/more-qr-use-examples/#

What technologies back up the QR Codes?

Smart phones, including iPhone, Nokia N series, Blackberry, Okta, Samsung, and LG, either come with the *reader* installed, or support the installation. Installs can be made from numerous websites. Google searches provide extensive lists of these services.

QR Codes can be generated through many online services, and the code is universal; another available option is leasing, much like a domain name.

Small Business Usage

Is there an ideal industry for QR Codes?

It can work for all industries. You may already have seen QR Codes enter the market through **Real Estate.** This is largely because Real Estate Agents have signage.

The QR Code codes can be placed on these signs with little additional cost, and can be specific to the property they advertise; i.e. a passer-by can walk past the property for sale at night, scan the code, and immediately be taken to a website that lets them see everything of interest inside the property.

Alternatively, the QR Code can direct the potential purchaser to the Real Estate Agency's general listing or to similar property listings. Similar scenarios can be achieved for most businesses.

What line of business are you in? Do you have signage, vehicles, bill boards, staff clothing, advertisements, or leaflets? Where could you put QR Codes? *(hint = any place you have your logo)*

How can QR Codes generate business for me?

Think of your market being **on the move . . .** What do they need to know? Where are they? How can you add value to them at that moment, right there?

Here's a good example:

Let's say you're a builder, your clients are mostly aged between 25 and 40, and local to your county or region, and they might want anything from minor repair and renovation works to the design and build of a brand new home. When you are working at a property, your sign-bearing van is there. You also have a small sandwich board sign in the front yard that has your contact details and a polite message that says *"we are carrying out renovation and building work, we aim to keep the noise to a minimum, please contact us onsite or online if you have any concerns about noise/dirt & dust."*

On this very polite sign, you have a QR Code image; this enables people to scan the code and be taken to your Internet or mobile website, (website optimised for mobile technologies) or even your mobile phone where they are able to comment and place feedback. This demonstration of courtesy and awareness speaks volumes for your dedication to service.

The neighbours, who are mostly your target demographic, may be thinking about renovations and would be more likely to consider using you. By scanning the code to get to your website, they are able to view the range of your previous work, your testimonials, and your credentials! Your website also makes it very easy to make an enquiry! It makes the whole process immediate and simple!

What have you changed in your business? Only a sign, a QR Code, and possibly added a mobile website - but what a different client experience!

Your staff members benefit too. On some machinery and tools you have placed a QR code label. These images take your staff to specific pages on your website. These pages have clear instructions on how to use the tools safely. This saves the cost of printed manuals which require replacement, and also ensures that the latest updates are always available. Your websites (both office-based and mobile) work as a huge resource for your staff. Your time is freed up and your staff members are productively engaged in their work as they are trusted and left to do their job!

Once you have finished the renovations, you are able to leave small QR Code stickers in relevant locations in the house such as on the hot water heater as a contact point should it ever fail or require assistance. The client can scan the code and immediately see an emergency number to ring, or be taken to the page on your websites that detail the plumbing work you do and the support you offer. This could also be done on the fuse box where you could provide basic information to help the client diagnose the fault themselves — to save them money and make them even happier.

And what have you changed in your business? Even now, only a sign, a QR Code, some sticky labels, and possibly some information added to your website . . . But what a different client experience!

Big Business Usage

Large businesses like Google, McDonald's, Coca Cola, and Audi, as well as individuals, families, bands, teachers, schools, artists, and more have adopted QR codes. See more videos, photos, stories, news, and ideas at the QRickit Blog

As we mentioned, QR codes are VERY BIG in Japan? You see them on almost everything — from soft drink cans and candy bar wrappers to posters, real estate signs, billboards, business cards, magazines, newspapers, and more. Well, this is not surprising since the QR Code was developed by a Japanese company, Denso Wave, in 1994.

In Japan, more than 90% of all music downloads are from mobile phones while less than 10% are from computers. Almost everyone has and uses a mobile phone or other handheld Internet device daily.

Now, Google USA has adopted the QR Code and is promoting its use in North America. Google recently sent thousands of free QR Code decals to selected businesses across the

USA to help bring customers to their mobile version.

In New Zealand, VodafoneNZ just used a QR codes in their latest mailout and tipped them as the next big thing in marketing.

Late in 2010, New Zealand's Lion Nathan Wine Group, owner and importer of fine wines from around the world, launched a new QR Code-based proprietary marketing program called Cellar Key.

Air New Zealand is using QR codes in their ticketing to speed check-ins.

There are limitless applications for QR Codes in big business. The QR Codes can be physically attached to products, inserted in printed advertisements, and placed on all forms of marketing materials whether paper-based or other surfaces such as uniforms, hats, vans, and vehicles, etc.

What could your message be?

First and foremost, it should include contact information that brings new customers to your website, mobile website, and phone. On a product for established customers, why not a quick re-ordering link? This will help you keep the client as it makes it easy for them to continue using you.

How about a feedback and reporting system? Allow your clients to directly provide feedback by scanning a code from your product and entering their comments on your specially provided website?

Benefits to the consumer

If you're wondering about the advantage over pen and paper, how many times have you taken down an e-mail address, website, or phone number with every intention of obtaining information for future contact — but then just never got around to it — or worse yet, simply lost the note? QR Codes enable the user to scan the code, open the link, and research the information *immediately*. Additionally, the link is saved for future access to the information.

Immediacy

The NOW generation: Generation Y is often referred to as the NOW generation. Generation Ys process information quickly and brutally. Their expectation is that the information presented will be relevant and directly answer the question they have at the moment. With properly structured and maintained websites, a scanned QR Code will *immediately* take the viewer to the relevant information required. Scanning a code eliminates any potential input errors, false and/or inaccurate search terms, and will immediately provide the information which is sought.

Some important features of QR Codes

High Capacity Encoding of Data

While conventional bar codes are capable of storing a maximum of approximately 20 digits, the QR Code is capable of handling **several dozen to several hundred times more information.**

QR Code is also capable of handling all types of data, such as numeric and alphabetic characters, Kanji, Kana, Hiragana, symbols, binary, and control codes. Up to 7,089 characters can be encoded in one symbol.

```
ABCDEFGHIJKLMNOPQRSTUVWXYZABCD
EFGHIJKLMNOPQRSTUVWXYZABCDEFGH
IJKLMNOPQRSTUVWXYZ012345678901
23456709012345670901234567090l
23456789ABCDEFGHIJKLMNOPQRSTUV
WXYZABCDEFGHIJKLMNOPQRSTUVWXYZ
ABCDEFGHIJKLMNOPQRSTUVWXYZ0123
4587890123456789012345678901 23
4567890123456789ABCDEFGHIJKLNN
OPQRSTUVWXYZABCDEFGHIJKLMNOPQR
```

A QR Code symbol of this size can encode 300 alphanumeric characters.

Small Printout Size

Since QR Code carries information both horizontally and vertically, QR Code is capable of encoding the same amount of data in approximately one-tenth the space of a traditional bar code. (For a smaller printout size, Micro QR Code is available.)

Dirt and Damage Resistant

QR Code has error correction capability. Data can be restored even if the symbol is partially dirty or damaged. A maximum 30% of code words*1 can be restored*2.

*1: A code word is a unit that constructs the data area. In the case of QR Code, one code word is equal to 8 bits.
*2: Data restoration may not be fully performed depending on the amount of dirt or damage.

Readable from any direction in 360°

QR Code is capable of 360 degree (omni-directional), high speed reading. QR Code accomplishes this task through position detection patterns located at the three corners of the symbol. These *position detection patterns* guarantee stable high-speed reading, circumventing the negative effects of background interference.

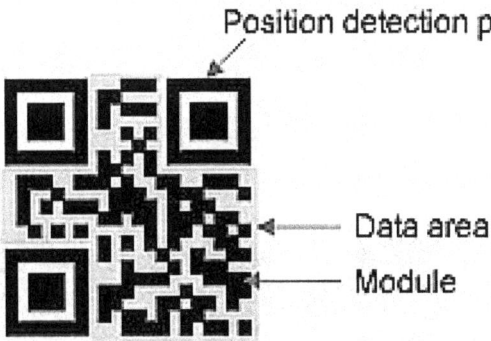

Position detection patterns

Data area

Module

Structured Append Feature

QR Code can be divided into multiple data areas. Conversely, information stored in multiple QR Code symbols can be reconstructed as single data symbols.

One data symbol can be divided into up to 16 symbols, allowing *printing in a narrow area.*

The same data can be read from the single symbol or from the four symbols.

Where is more information available?

The Internet is the best resource for more information concerning QR Codes. If you want to know more, you can scan this code.

For assistance in customizing your own QR Codes, you can contact qrcodequeenz@gmail.com

In a nutshell

QR Codes can hold over 200 times the data of a regular bar code and have more than a 30-fold increase in recording density.

QR codes have 360 degree multidirectional, high-speed reading: no need to align the reader to the code, simply point and scan.

They are dirt and damage resistant: data redundancy ensures that the code can be read if part of the code is missing or damaged

They have warp resistance: code can be placed on a curved or warped surface.

Different language formats can be encoded: supports other languages' character sets such as Japanese, Chinese, etc.

High capacity, space-saving printing uses less space, less ink, and less intrusion into valuable marketing display space.

Linking function: one QR Code can be divided into 12 QR Codes.

How much does it cost?

Scanning the codes costs nothing; however, since some codes have a web link to prompt mobile browsing, you will need to confirm access cost with your local communications provider. Also, while several websites offer free codes, these do carry the provider's name as well. Obtaining special, customized codes may incur a small expense.

What is the next step for your business?

Get started today!

And the QR Codes
are not just for BUSINESS!

Magazines, billboards, and even product packaging — including music and movie CDs and DVDs — have used QR codes to deliver enhanced content to consumers who scan them with a phone camera. While true AR applications overlay digital information on a view of the physical world through the camera's lens, QR codes simply link to that information. In many cases, the code contains a URL that, when scanned, sends the phone's browser to a website. Additional data such as contact information or even an e-mail message can also be embedded in a QR code.

QR codes are still somewhat a novelty in the Western world, but in Japan they are already a tried-and-true method for sharing data. Widespread smartphone adoption in that country has made QR code access more widespread.

So what else is being done with the technology?

Books

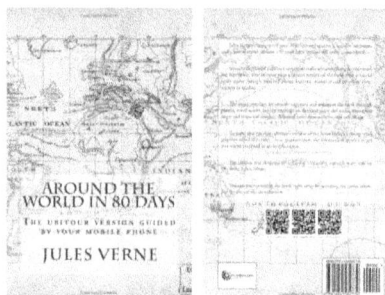

Ubimark Books is using QR codes as it pioneers this exciting connectivity device. Its enhanced version of _Around the World in 80 Days_, for example, contains QR codes directing readers to online audio, video, maps, and other interactive features. Ubimark have republished Jules Verne's classic _Around the World in 80 Days_ enhanced with QR Codes. The QR Codes, two to a page, resolve to a mobile site where readers can participate in online conversations about the book, follow Phileas Fogg's adventure on interactive maps, and play audio or video versions.

For example the QR Code at the beginning of Chapter Two links to this page.

To see more, visit
http://2d-code.co.uk/around-the-world-with-qr-codes/

This particular application niche especially appeals to me as a copyeditor who works with both fiction and non-fiction text in today's world of digital publishing. The potential for immediate connectivity and interaction with the vast resources available online is truly mind-boggling; I can't wait to use it in my own material and to enrich my clients manuscripts! For that reason, several QR codes have been included in this text so you, the reader, can try it out for yourself.

Libraries

QR codes can be used to deliver a higher level of support and interactivity to patrons. Even better, the technology is a snap to implement — at little or no cost.

Among several sites that let you create codes for free, QR Stuff lets you make one that includes URLs, e-mail, and phone information. And even more dynamic codes are possible, ones that automatically send an e-mail or SMS message. A school could implement an e-mail or SMS reference service via QR codes placed around campus. Students with questions could then initiate a reference contact by scanning the QR code with their phones.

Schools

The 2009 Australian and New Zealand Horizon Report places QR codes on the horizon for new technologies that will be adopted in the next four to five years. It is interesting to think of a technology that you are only just meeting actually being in common use in such a relatively short time frame.

A veritable encyclopedia of educational potential for QR codes can be found at http://livebinders.com/play/play_or_edit/52126 .

Look at the range of data that can be encoded . . .

- Website URL
- Text
- Telephone Number
- SMS Message
- E-mail Address
- E-mail message
- Contact Details (Vcard)
- Event)V Calendar)
- Google Maps Location
- WiFi Login (Android only)
- PayPal BuyNow Link
- Social Medial
- iTunes Link
- YouTube Video

Cleaner and Greener

The U.S. Environmental Protection Agency is proposing a new car label that incorporates a QR code with fuel economy information, comparing the scanned car against other models.

NOTES

QR Code Reader Software

In order to take part in this new dimension, your mobile phone must be able to read the QR Codes. Check to see if you already have a reader on your phone (readers are also available for installation as an app on your PC).

If you need to add a reader, check out one of the following sites; they offer a range of options such as downloading via PC and then transferring to your phone, or visiting with your smartphone via browser for direct download and installation. Some also have the option of setting the application up on your phone as a device bookmark.

You can also check with your mobile phone vendor or visit a local App Store.

- **Optiscan** - The best QR Code scanner for iPhones - it understands all the new trickier encoding types and has definitely kept up with development in QR code technology.
- **I-Nigma** - Probably the most popular decoder/reader application and works on

most of the popular smartphones. (Supported Devices)

- **Quickmark** - For most phones, but most noticeably has an Android version and a Windows Mobile version
- **Barcode Scanner** - Another good Android QR code reader. Available in the Android Market in the Applications/Shopping category.
- **Kaywa Reader**
- **Nokia** - Most Nokia's now come standard with the Nokia's own scanning software, but this one is good for N78, 6210 Navigator, N80, N96 and 6220 Classic
- **Google Zxing** - For the Android and iPhone plus quite a few others, but not Windows Mobile
- **SnapMaze** - QR code reader for Nokia, Sony Ericcson and Motorola phones (Supported Devices)
- **NeoReader** - A good range of phones and mobile devices (including iPhone and Blackberry) and also available as AppStore download. (Supported Devices)
- **Jaxo Systems** - Runs on most Java-enabled phones (Supported Devices)

- **OkoTag** - The new one from Jaxo. Java, Blackberry, Windows Mobile and Android.
- **Blackberry Messenger** - Comes standard with most Blackberrys these days
- **Upcode** - Support for a wide range of Symbian, UIQ, windows mobile, iPhone, Blackberry, and Java phones.

QR Code Software & Applications

- Bar Capture - Capture and decode QR Codes from your computer screen
- Online Decoder - Decode QR Codes online via direct image file URL or image download
- Wordpress QR Code Plug-In
- InDesign CS3 QR Code Plug-In
- Firefox Mobile Barcode Add-On
- Facebook QR Codes application
- QR-Code Tag - Google Chrome QR Code Plug-In

MORE IDEAS
FOR USING QR CODES

Here's where the magic comes in; as we get our heads – and imaginations — around this new technological paradigm, it seems that the virtual magic of this immediate interactive connectivity — held in the palm of our hands — has no limits. Once we enter, it goes on and on and on . . .

Over the past year, excitement about QR Codes has grown seemingly on a daily basis. We've seen more and more businesses use them to promote their company, products, or services. We've seen libraries and schools put them to use to help students and parents. We've also seen them provide value to specific industries such as real estate and healthcare.

A great blog that demonstrates this is *Qreate and Track*:

http://qreateandtrack.com/2012/01/

Here's a list of 100 ways that you could possibly use QR Codes. We hope that you find at least a few of them helpful.

100 Possible Ways
to Use QR Codes

- Put them on signs, so people can track bus routes and arrival times

- Add them on/near historical monuments, direct them to videos with more information

- Put them on city-owned vehicles so people can see how much it costs and how quickly it's depreciating

- In the "Autos" section of a newspaper/flyer/magazine, put a QR Code for people to see the Autos of the Week

- Add a QR Code on promotional posters that upon scan, allow people to "Like" pages on Facebook

- Put QR Codes on clothing to drive people to your Twitter page

- In photo-books, put QR Codes that drive people to the LinkedIn Profiles of people in the photos

- In the Real Estate section of a newspaper/flyer/magazine, put a QR Code for the House of the Week

- In real estate, add a QR Code to sign riders. Direct people to a video/virtual tour of the house.

- Add a QR Code on your business card that provides people with your contact information

- Add a QR Code on your business card that directs people to online site – it might contain a video of you, links to social media profiles, and your corporate website.

- Add QR Codes to various sections of a book that direct people to enhanced content on the web.

- Put a QR Code next to columnists in a newspaper. It can direct people to an online page that lists other articles from that writer.

- In restaurants, put QR Codes on the table tents. These may direct people to social media pages for the restaurant.

- In stores, add QR Codes on windows, cash registers, and other signs. These may direct people to mobile-optimized pages where they can sign up for your newsletter.

- In your company brochures, add QR Codes that direct people to videos, podcasts, and web pages that further discuss the corresponding topic from the printed piece.

- Put QR Codes on Direct Mail pieces that drive people to personalized URLs.

- Add a QR Code to a print advertisement for a restaurant. Upon scanning the code, direct people to a page where they can make reservations.

- Hotels could put QR Codes in places that direct people to mobile-optimized sites where they can make reservations.

- Running an advertisement in multiple magazines? Create a unique QR Code for each magazine. Find out which one is driving the most traffic for you.

- Make it easy for people to shop: Put a QR Code on a direct mail piece for a retailer. Drive people to a mobile-optimized site that allows them to browse and purchase.

- Does your store provide daily specials online? Allow people to scan a QR Code that directs them there. They can easily return to that URL each day from their barcode scanning app.

- In the construction business? Put a QR Code on the sign you place outside recent projects. Upon scanning the code, people can link to more information about your company.

- Printing a guide for a film festival? Add QR Codes that allow people to see trailers for the films.

- Promote your company's social media pages with QR Codes. Have a Facebook page? Allow people to scan and then be instantly directed to it on their phone.

- QR Codes can be scanned from a TV. Display one during a news broadcast to drive people to your online content.

- Make it easy for people to see your customer testimonials – add a QR Code to your next newsletter that drives people to an online page of written or recorded testimonials.

- Further expose people to your brand — add your logo to the middle of a QR Code for additional exposure.

- Make your weekly flyer in the Sunday newspaper interactive – allow people to scan QR Codes to easily learn more (or place an order!) for your featured products.

- Pay for stuff! QR Codes can be directed to sites that will automatically deduct money from your account for a purchase.

- Make it easy for people to assemble your products. Put a QR Code on the box or on the directions that allow them to see videos that show how to put something together.

- Trying to promote a band? Add a QR Code to their poster that lets people view tour dates, videos, and more.

- Get the most out of your promotional items. If someone's going to keep a coffee mug or calendar with your logo on their desk, add a QR Code so they can easily access your company's website.

- Promoting a movie? Put a QR Code on the poster to allow people to see showtimes in their local area.

- Make it easy for people to find your business. Put a QR Code in one of your print ads that connects people to Google Maps.

- Provide incentive for people to travel to your resort. Add QR Codes that link people to video tours of what you offer.

- Reward people for buying your product. Add a QR Code to the label to drive people to online community sites.

- Make it easy for people to re-order your products. Put QR Codes on labels, boxes, and other packaging materials that direct them to online shopping functions.

- Put a QR Code on the back of a concert ticket. People often save those. A QR Code will allow them to instantly access more information about a performer.

- Give people something to do while they're waiting. If you often have a captive audience that is waiting in line, put a QR Code on a nearby poster or billboard. This will allow them to interact with your brand while they are waiting.

- Put QR Codes on billboards on the side of the road. and near parking lots, scanning the code will give more information about your business.

- Drive donations for fundraising efforts... Make it easy as possible for people to give by adding a QR Code that drives them to a contribution page.

- Make your trade show booth interactive! Put QR Codes on panels and other materials so that visitors can scan a code and interact with your brand even after the show is over.

- Increase registrations for a contest — allow people to scan a QR Code that drives them to a mobile-optimized registration page.

- Add a QR Code to a poster in the library that allows people to read reviews of featured books.

- Treasure/Scavenger hunts! For a bit of fun, add QR Codes throughout a certain area (indoors or outdoors). Codes can direct people to clues as to where they go next to finish the hunt.

- Put a QR Code at the bottom of your blog posts. When people print and save them, they'll be able to easily get back to your blog later on.

- Inside a classroom, put QR Codes to direct students to the school's YouTube channel.

- Ask a Trivia Question. The QR Code can lead people to a simple form where they can answer to win.

- Put QR Codes on a globe or map to allow people to access additional information about a specific location.

- Make memories last for a long time – put QR Codes in a yearbook that may direct people to a Facebook page or community website about their class.

- Get someone to buy that DVD; put a QR Code on the case that lets someone watch the trailer.

- Recommend Further Reading; put a QR Code at the end of a book to drive people to a site where they can learn about related books and topics.

- For schools, allow parents to meet your staff. Put QR Codes in the directory so parents can access more information about a specific teacher or administrator.

- Show people just how fresh your product is; add a QR Code to the packages for produce to allow people to view the "born-on" date and location.

- Help people select the best beverage; add a QR Code to wine bottle labels to give them a virtual tour of your winery.

- Make a recipe book interactive: add a QR Code to each recipe to show a video of the dish being prepared.

- Bring your music-related product to life; add a QR Code to printed materials so people can hear and/or see how an instrument will benefit them.

- Stand out from the crowd; wearing clothing with a QR Code on it will certainly help draw attention; and depending on where it points to, it can help you build your brand.

- Develop interest in your college/ university by putting QR Codes on printed materials that link people to virtual campus tours.

- Help students choose which classes they should take. Use QR Codes to allow them to access interactive information about a course, to help them better make an informed decision.

- Include QR Codes on homework to provide supplemental material for students.

- For schools, add QR Codes on printed materials sent out to parents or guardians to make it easy for them to either gain additional information, or to contact the appropriate personnel.

- Promote exercise and other athletic endeavors. Want to show someone how to properly play volleyball? Put a QR Code on a poster to link to a video demonstration.

- Make it easy for someone to check-in to your business. Put a QR Code outside your store. Scanning it allows someone to easily "check-in" via a site such as Foursquare.

- Encourage user-generated content for the news. Put QR Codes throughout your city or town. When people are walking around, they can scan the code to access an easy-to-use "submit your news tip" to a newspaper, radio station, TV station, or a community website.

- Add them to badges that conference or trade show attendees wear. This will make it easy for everyone to share their information with others.

- Hosting an outdoor festival? Put QR Codes on signs and posters. They may direct people to a mobile-optimized festival guide and schedule.

- Promote school teams and groups. Include QR Codes on related printed materials so people can view online information about schedule, participants, history, coaches, teachers, etc.

- Add a bit of excitement to the Summer Reading List for students: add a QR Code next to each book so students can learn more before they read.

- In-store: Provide more information about products. Allow potential consumers to read online reviews.

- Promote local businesses – the Chamber of Commerce could include QR Codes for its members on printed materials, so that residents can easily learn more about ones they are interested in.

- Make the periodic table of elements interactive; include QR Codes that lead to more information on each element.

- Help students make better use of school equipment. For example, put QR Codes on tools in the science lab that provide tutorials on proper usage, along with ideas on why they should use the tool.

- Enhance your resume with a QR Code! Allow potential employers to learn more about you online through videos, or links to projects that you've worked on in the past.

- Provide more information about plants and flowers... A QR Code can let people see what their purchase may turn into, along with best practices to help it grow successfully.

- Allow people to view the "making of", or a behind-the-scenes look at how your team built a product.

- Expand your science fair exhibit. Participants typically have a limited amount of space. QR Codes may allow attendees to learn more about your project, and what you learned in the process.

- Easily share audio — create QR Codes that drive people to MP3s for songs or to podcasts.

- Provide QR Codes that link to a class' blog. This may be included on homework or other notes. It will direct students to more information when they need help.

- Allow parents to learn more about school trips... Include QR Codes so that parents can see videos about where and for what purpose students may benefit from participating.

- Enable people to access additional training materials while perusing through a manual or guidebook.

- Speaking or presenting at a conference? Include a QR Code in your handout so that people can learn about you, your company, and the topic.

- Provide additional instructions for physical therapy exercises. In addition to listing steps on printed material, allow people to gain more information online.

- Help patients visualize procedures and surgeries. QR Codes can link to videos to provide some further detail as to what's involved.

- Allow people to interact with your greeting cards. Sure, people may love to hear a note of Congratulations or Thank You. A QR Code can allow them to view a personal picture or video from you that also echoes those greetings.

- Connect your printed newsletter with online material; enable people to easily access additional information about a topic online.

- Enhance your customer service offerings. Next time you distribute release notes or training manuals, include QR Codes to link people to online tutorials or ways to reach your support team via their mobile phone.

- Add a bit of fun to your company's Open House; display signs and banners with QR Codes that enable visitors to enter contests, or learn more about company projects.

- Allow people to "cut the line"! Put a QR Code on tickets or boarding passes that make it easy for them to have their pass scanned and move to their next destination.

- Enable people to access additional information from bills and invoices. QR Codes can direct them to an online site where they can view information about your products and services, as well as learn about what else you offer.

- Promote future events! If you do not want to devote a lot of space on a printed item to promote an event that will occur in the future, you can include a small QR Code that directs people online to learn more.

- Enhance your brand by providing free tips: allow a prospect to scan a QR Code on a piece of clothing that you are selling. It can direct them to pictures or fashion tips as to how they could put that item to use.

- Give more legs to everyday items. There are things that people buy every day: coffee, soda, water. Put QR Codes on labels for those products to help engage your loyal consumers even more.

- Provide nutritional information on menus to help people decide if something is right for them to order.

- Build your marketing database; include QR Codes on your printed materials to enable people to easily subscribe for e-mail or SMS alerts.

- For fun and games, have people scan a QR Code that takes them to a landing page. Include "randomizer" logic on the page that tells them if they are a lucky winner.

- For schools – keep parents in the loop about their child's activity. Include a QR Code on homework assignments. Scan the returned ones in the next day. If a child's homework is not returned, have a notice sent to parents instantly.

- Measure what locations are responding to a product; put QR Codes on products that have a wide distribution. Track where the scans are coming from to see where your hot and cold spots are.

- Put QR Codes on boxes of food-related items that lead people to recipes that could utilize that ingredient.

With special thanks to http://qreateandtrack.com/2010/12/22/100-ideas-on-how-to-use-qr-codes/ for this wonderful collection of ideas.

References and Links

- http://www.washingtonpost.com/wp-dyn/content/article/2010/09/07/AR2010090706625.html
- http://qreateandtrack.com/2010/12/22/100-ideas-on-how-to-use-qr-codes/
- http://www.youtube.com/watch?v=5RFQ0cnSOhg
- http://www.qrmonkey.com/more-qr-use-examples/#
- http://qrickit.blogspot.com/
- http://www.denso-wave.com/en/adcd/
- http://www.google.com/help/maps/favoriteplaces/business/barcode.html
- U.S. Environmental Protection Agency
- http://2d-code.co.uk/around-the-world-with-qr-codes/
- http://ubimark.com
- http://www.qrstuff.com/
- 2009 Australian and New Zealand Horizon Report
- http://livebinders.com/play/play_or_edit/52126
- http://www.qrstuff.com/
- http://www.proofreadnz.co.nz
- http://www.qrcodequeenz.com

POST SCRIPT

In stumbling upon the new world of the QR Code 'dimension', at first it was quite a project to find material that would help me to learn more about it . . . and when and how to make effective use of this incredible development.

It continues to be an exciting and delightful adventure that has inspired me to collect some of the best references and resources and put them into this little book in the hopes that it will speed your advance into this new paradigm.

My profound thanks and deepest appreciation to those who are blazing the trail and who have provided much of the information I pass on. I hope the readers will visit your websites and make use of the videos, tools, and services offered.

Scan you soon,

Judith Sansweet
www.qrcodequeenz.com

MY QR CODE IDEAS

MY QR CODE IDEAS

www.ingramcontent.com/pod-product-compliance
Lightning Source LLC
Chambersburg PA
CBHW050530210326
41520CB00012B/2520